SPLASH and TRICKLE

A Conservation Story

written by

Ivah Green

illustrated by

Bil Connor

dedicated to

Ann

published by

ODDO PUBLISHING

Mankato, Minnesota

Library of Congress Catalog Card Number 64-25750

Printed in the United States of America.

One hot summer day a big black cloud appeared in the sky. Soon millions of raindrops began falling.

One of those raindrops was SPLASH.
Here he is,

Another raindrop was TRICKLE.
He's over here.

SPLASH and TRICKLE were excited, for they were on their way to the earth below. Where would they fall, they wondered. What would happen to them on the earth? They thought:

Shall we be together?
Shall we reach the sea?
What a BIG ADVENTURE
Ours is going to be!

Just then their BIG ADVENTURE really began. A gust of wind came along. It whisked SPLASH in one direction. And it tumbled TRICKLE in another direction.

"Hey!" called out TRICKLE to SPLASH. He was a little breathless, for the wind had whirled him through a somersault. "Maybe we won't be able to stay together. You go your way, SPLASH, and I'll go mine. Hope you have a pleasant journey!"

SPLASH waved goodbye to his friend. "Some day I'll get back to the sky and see TRICKLE again," he thought. "But all sorts of things may happen to me before that time."

The wind tossed SPLASH this way and that—and now he was very close to earth. He thought:

"Wonder where I'm going—
Wonder what I'll see;
Guess I'll try that hillside;
It looks good to me."

If SPLASH had expected to find a soft, smooth, grassy bed to land on, how disappointed he must have been. Instead, he saw only a bare hillside—no grass, no clover, no trees. Just hard, stony ground. When he struck the earth he felt as if he had been splattered into a hundred pieces.

"Oh, my goodness!" SPLASH exclaimed,

"How I splash!
And how I splatter!
No grass here,
That's what's the matter!"

SPLASH had been more surprised than hurt, but he pulled himself together. Then, before he could cry "Help!" he was lifted right off his feet and hurled pell-mell down the hill.

That wasn't what he wanted to happen, at all. So he pushed and he pulled, trying to stop himself. He dug . . .

First with the heel,
And then with the toe,
But into the ground
He could not go.

The side of the hill was so rough and hard and stony that he couldn't get a toehold in it. Down, down the hill he rolled, faster and faster. He mumbled as he rolled:

"*Roll away, roll away,*
Down the hill;
Nothing to stop me—
Can't stand still."

"Sooner or later I must stop," he thought, then wondered:

"*What shall I find waiting down below the hill?*
Wish I'd meet some clover, that would hold me still!"

By this time other raindrops had joined SPLASH. They, too, were tumbling in a mad rush toward the bottom of the hill.

No green grasses to hold them,
No cushion-y patches of clover,
They turned somersaults down the slope
Over and over and over.

Soon they came to a deep furrow, like a ditch. The farmer had made that furrow and many others when he plowed up and down the straight sides of the hill. Every time the raindrops fell on this hillside they had taken those furrows for their path down. Now there were deep gullies where furrows had once been.

SPLASH murmured sadly:

"Raindrops should be helping
Grass and grain to grow;
How we'll do it this way,
I should like to know!"

Yes, SPLASH had hoped that on his BIG ADVENTURE he would give a drink to a blade of grass, or a leaf of clover, or a tiny flower. "Then I could feel I was doing something important," he thought. "That's what raindrops are for— to help things grow. Yet here I am, swirling down a gully! Goodness only knows where I'll finally land."

Just then SPLASH noticed that something new and strange was happening to him. His face was getting muddy. Mud was filling his eyes and his ears. The mud was soil from the hillside and the sides of the gully. The raindrops around him, too, were slowly being covered with mud. Was this the end of SPLASH's plans for helping something to grow?

"All my dreams," cried SPLASH,
"Now nipped in the bud;
I can see my finish—
Just a ball of mud!"

But SPLASH didn't turn into a ball of mud. As more raindrops splashed near him, the mud dropped away from him. Now he could see better. He discovered that he and his raindrop companions were part of a swiftly moving river— a flood! They were rushing over beautiful farms, pouring into houses, barns, chicken coops and hog pens.

Poor, frightened farm families! They tried to save their furniture, their horses, their cows and chickens. They had to be taken away from their homes in boats.
Horrified, SPLASH watched men and boys working with bags of sand along the river's edge, trying to keep it from flooding the land. He saw uprooted trees float past him as he swept along on the roaring water. For days and nights the swollen flood waters foamed and splashed and roared past farms and schools and country towns.

SPLASH thought sadly:

"Such a SAD ADVENTURE!
Nothing like I planned!
We are stealing—stealing—
Some good farmers' land!"

SPLASH was right. The flood was carrying away tons and tons of the rich soil on which farmers had expected to raise corn and oats and wheat and rye. The flood was a robber!

"Who is to blame for this terrible flood?" SPLASH wondered. Then he remembered the furrows that had started the ditches on the hillside.

The flood waters pushed the soil along until all the streams finally reached the ocean. There the soil drifted down through the water and came to rest on the bottom.

This was a sad thing to have happen, because many people in the world were hungry. And the good rich soil that might have helped to grow food for them was now beyond reach of the farmers. No corn or oats or wheat or rye could ever grow on the soil at the bottom of the ocean.

SPLASH felt sad.
"I never thought," he whimpered,
"When I fell today,
That my BIG ADVENTURE
Would turn out this way!"

"But here I am," SPLASH muttered to himself as he splashed about in the ocean. "If only I could have stayed longer on the land back there! Maybe I could have helped some plant to grow. I could have sprinkled it and given it a good drink. Then it might have grown to be big and strong. But now . . .

It's too late for wishing,
For the soil is gone;
And the mighty river
Still goes roaring on.

"If I and the other raindrops had traveled more slowly as we rolled down the hill," SPLASH went on, "we might

not have helped to make a flood. How nice it would have been if all of us could have given grass and clover and alfalfa and grain a drink instead. Now I must stay here in the ocean for a time, along with thousands of others like me. But someday the kind old sun will draw us back up into the sky. I'll float around in a fleecy white cloud for a time. Then some cool gray day I'll fall to earth again."

But where was TRICKLE all this time? Had he, too, been disappointed in his BIG ADVENTURE? No, for to his delight he had fallen on the broad green blade of a grass plant. As he dropped upon it, the blade bent down toward the ground, and TRICKLE hardly knew when he stopped falling. He just slid quietly to the ground.

"Hi, there," said the grass blades,
"Glad you came our way,
We would like to visit,
If you've time to stay."

"Hello," answered TRICKLE. "I'm just as glad to see *you*. Here's a happy beginning to my earth visit. I was afraid I might hit the ground with a hard bump. But this is the kind of soft landing place I like."

"We like it here, too," said a grass blade. "Do you notice how soft the ground is all around you? The raindrops first strike our long, drooping leaf blades. Then they roll off onto the ground the way you did. They don't hit the ground with a bump and a splash. The soil doesn't wash away, either, around us. It *stays in place*."

"I'd like to stay," said TRICKLE, "but I have work to do. Goodbye." He rolled along slowly and easily on the sloping land. "I'm surely having an easy time on my BIG ADVENTURE," he thought. "Wonder where I'll find some thirsty plants."

At that moment TRICKLE stopped moving. A big ridge of soil rose right up in front of him. He knew he could not climb over it. Could he travel around it? No, for as far as he could see in both directions the ridge of soil went on and on.

"Ho! Ho! Little Raindrop,"
Laughed a voice close by;
"You cannot get over
No matter how you try."

A tall corn plant was talking to TRICKLE. "This ridge was put here when the farmer plowed around the hill," it said. "Here the furrows and the corn plants run on the curve, or contour, of the land instead of up and down the hill. If the ridges weren't here, raindrops would be rushing past us, and stealing the soil we grow in.

This farmer is wise. He calls himself a conservationist. He says he intends to keep the raindrops where they fall. He isn't going to let his good rich topsoil be washed downhill and into the ocean."

"That's fine," replied TRICKLE. "I wouldn't want to steal any soil from a farmer and his corn plants. So, if I can't get *over* this ridge, I'll try going *under* it." And he began working his way down into the soil near the cornstalks.

"This is just what we raindrops like," he thought. "We want to trickle into the soil, and drop slowly from the top of the land to the roots of plants. Then we can do our best work." And he made up a little song as he kept sliding down into the soil.

"Trickle and drip,
Trickle and drip,
Into the soil
We'll quietly slip.
Quiet and slow,
Quiet and slow,
Down in the soil
We'll help plants to grow."

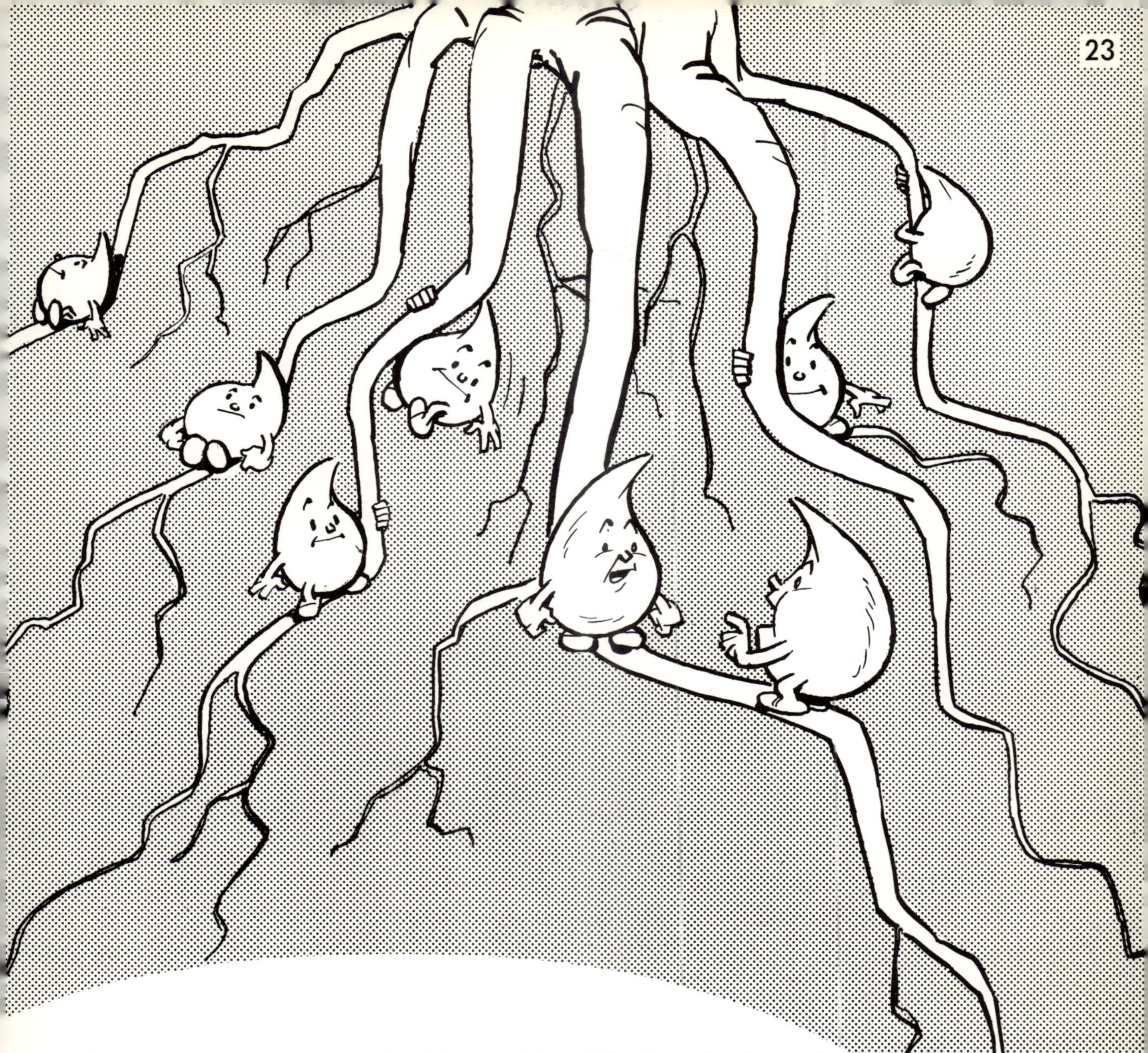

By this time TRICKLE had been joined by hundreds and hundreds of other raindrops. They, too, had been stopped in their journeying by ridges of soil. So they, too, were cheerfully sliding down into the soil. They had no trouble at all getting into it.

"What makes it so easy?" asked a raindrop close to TRICKLE.

"I think it must be because there are so many air spaces down here," replied TRICKLE. "Do you notice the roots and stems of these grass plants? They have room to move around down here. And there are dead roots and stems, too. They help to keep the soil soft, and they leave places for air and water to get in."

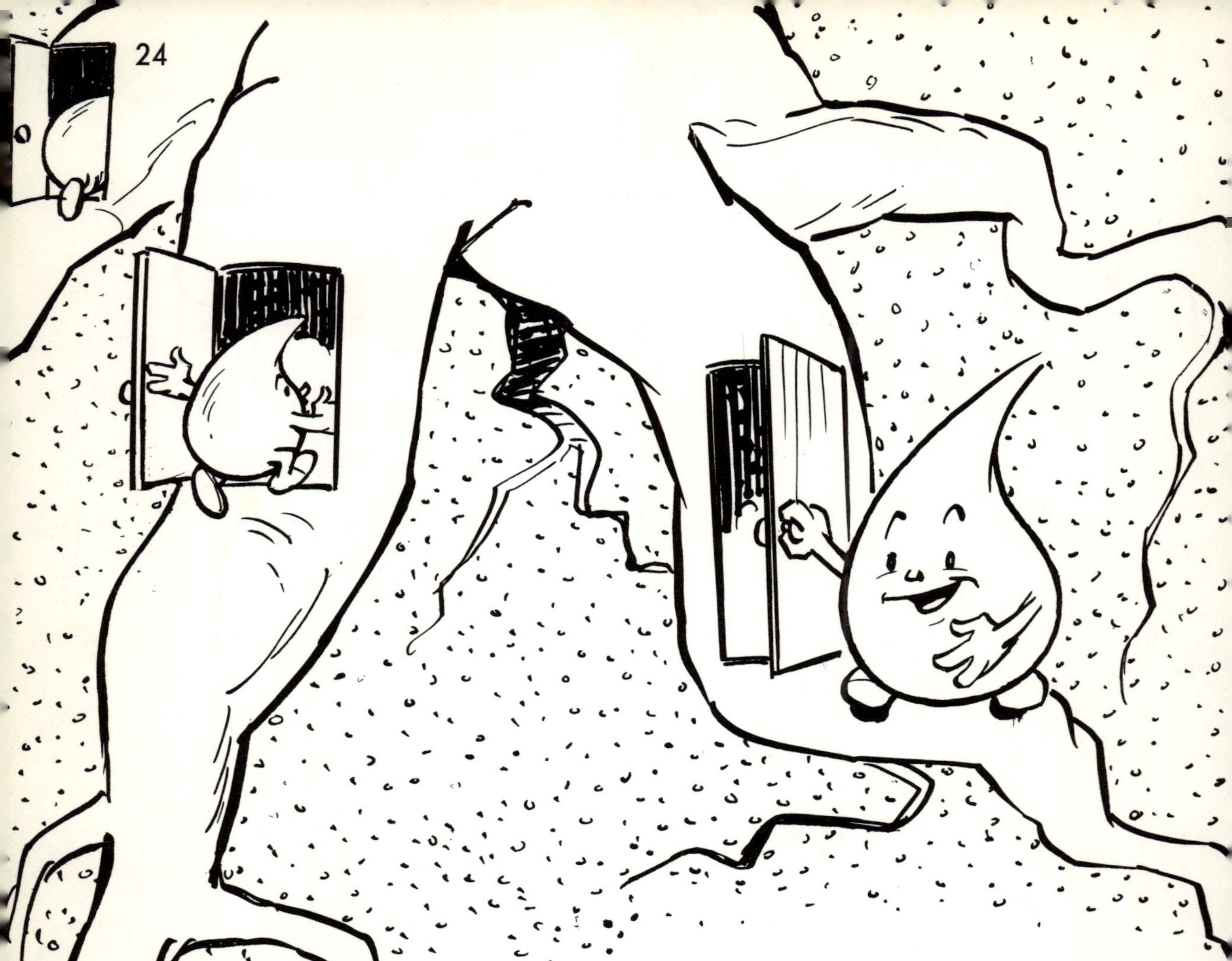

"It's a *live* root that I want to find," said TRICKLE's companion. "Here's a fine big one. I'll go inside and give it a drink."

Many other raindrops found roots where they could enter. TRICKLE watched them, but he had other plans. He told the raindrops around him:

"There is something calling,
And it says to me,
'Come and join the river
Flowing to the sea'."

TRICKLE and thousands of other raindrops had joined together to become a little stream, running through the roots and soil underground. He had not traveled very far before he had found the river.

Slowly and quietly they all moved on together, through the cool dark earth. Then after days and weeks of this quiet traveling they saw sunlight ahead.

At last they flowed out through an opening in the side of a hill. They were part of a spring.

The farmer had turned the spring into a pond for the use of his livestock.

"This isn't the sea I expected," thought TRICKLE, "but it's better, perhaps. Some of us will give drinks to cattle and horses. Some of us will spill over onto the grass below the pond. We'll help the grass grow rich and green. We'll supply water for fish to swim in, and plants to grow in. And some day the sun will draw us back into the sky."

TRICKLE felt needed, and important in the farm pond. Then one warm day he felt himself being pulled upward by a strong power that he could not see, but only feel. The sun was pulling him and thousands of his raindrop friends back up into the sky.

TRICKLE was glad to go, for he knew that he had finished his work for a time.

"My BIG ADVENTURE's over,"
He told the kindly sun;
"But send me down again soon,
I'd like another one."

One day when a big gray cloud formed in the sky, TRICKLE met his old friend SPLASH. They were glad to see each other again and they talked about their trips to the earth. TRICKLE was happy about his BIG ADVENTURE which had turned out to be a *glad* one, but SPLASH was unhappy about *his* adventure which had been a *sad* one.

Then as the cloud began to drop its heavy load of raindrops to the thirsty land below, TRICKLE and SPLASH fell toward the earth just as they had done before.

A farmer standing in his doorway, looked up at the sky. "It's going to rain, Mother," he called to his wife. "I hope it will be a good, slow, soaking rain. Just the kind my pastures and crops need. And my land is ready for it."

For that farmer was a conservationist. His farm was prepared to take the rain and put it to use. There were no gullies, because he had plowed on the curves of his land. He had left deep ridges of soil in just the right places to stop runaway raindrops. Grass and clover grew in the pastures. A farm pond fed by a sparkling spring offered water to everything that wanted a drink.

When TRICKLE and SPLASH dropped on that farm there was no bumping and no splattering. Just a delightful feeling of lightness and "springyness" on the cushion-y places of grass and clover and soft, rich soil.

Here the raindrops would work together to help the farmer. Here they would find their BIG ADVENTURE. Happily they chanted:

"Our BIG ADVENTURE
Will not be a sad one;
This time we're certain
We'll make it a glad one.

For farmers and raindrops
And sunshiny weather
Can make a good world
By all working together."

The end.

about the author

Ivah Green has earned her B.Ed. and M.A. degrees from St. Cloud State College and George Peabody College for Teachers, and has taught in various levels of education.

She has been a director of conservation workshops for elementary teachers and is the author of several books dealing with conservation and wildlife.

ACKNOWLEDGMENTS: I wish to express my appreciation to Bil Connor for his appealing illustrations, and to Morris Fonda who gave me the idea for the beginning of this story.

about the illustrator

Bil Connor was born in Elgin, Illinois and after his formal schooling, attended the Chicago Academy of Fine Arts.

His art background has taken him into all forms of illustration, working in Chicago Commercial Art Studios; Omaha Bee-News Art Department; in the Animation Department of the Walt Disney Studios in California; for more than twenty years in the Art Department of Des Moines Register and Tribune, and for the same paper on several comic strips in their Syndicate Department, (a separate organization within the Cowles enterprises).

He is married and the father of three sons.

In World War II he served in the Army for two years.

His hobbies consist of baseball, basketball, swimming, and bowling.

Reading Consultant— *William E. Jones, Lancaster Public Schools, New York*

the **WONDERFUL WORLD** of **CHILDREN'S READERS**